Tick Tock Telling Time
Time to the Hour and Half Hour

Kathleen L. Stone

Copyright © 2016 Kathleen L. Stone

All rights reserved.

ISBN-13: 978-1532966279
ISBN-10: 153296627X

Enjoy these other books by Kathleen L. Stone

Penguin Place Value
A Math Adventure

Number Line Fun
Solving Number Mysteries

Riley the Robot
An Input/Output Machine

Mason the Magician
Hundreds Chart Addition

Katelyn's Fair Share Picnic
More Math Fun

Money Tree Mysteries
Adventures with Quarters

Alien Even and Alien Odd
A Math Space Adventure

Kenley's Line Plot Graph
Another Math Adventure

Matthew's Sunshine Bakery
Multiplication Arrays

Firefighter Gary's Fire Safety Rules

Samantha's Search
3D Shapes

Grandma's Quilts
Fun with Fractions

Daniel's Day of Multiplication
Multiplication with Equal Groups

More Penguin Place Value
Hundreds, Tens, and Ones

Dedication

With all my love to my family ... time spent with family is the best time of all!

How many of you remember
The rhyme Hickory, Dickory, Dock
And that cute little mouse
That kept running up the clock?

Well that mouse has been busy
In fact, she's only just begun.
She's back again to teach us
Even more telling time fun.

Learning to tell time isn't hard.
It's as easy as one, two, three.
This little mouse will show us
Just how fun telling time can be.

When the big hand is on the *twelve*
And we hear the clock tick tock
We know that the time it's telling
Will always end with "*o'clock.*"

But our tiny mouse friend knows
It's the little hand that has the power.
The big hand counts the minutes
But the little hand tells the hour.

When you want to know the time
Always start with the little hand.
It tells you the hour to say,
Now isn't that grand?

So take a look at this clock
What time do you see?
Our mouse says *three o'clock*?
Thumbs up if you agree.

Here's another clock for you.
Its hour hand is on the *one*.
Did you say *one o'clock*?
I told you this would be fun!

Here's a cute flower clock
With its hour hand on the *nine*.
I bet you said *nine o'clock*.
You are really doing fine!

Let's have a little contest
And see who will win.
What time will it be
When the hour hand is on the *ten*?

I heard you say *ten o'clock* first.
Let's try just one more.
What time will it be
When the hour hand's on the *four*?

Tick Tock

Four o'clock? That's right!
Now what is the mouse going to do?
She's moving the minute hand
To teach us something new.

Let's have another contest
And see if you can win it.
As the big hand moves around
It will tell us each minute.

As it moves around the numbers
We will count by *five*.
When it's on the *two* it says "*ten*"
On the *eleven* it says "*fifty-five*."

4:30

When the minute hand is on the *six*
"*Thirty*" is what it will say.
But the hour hand gets tricky
When we are telling time this way.

The number the hour hand passes
Will tell us what time it will be.
Don't let that hour hand fool you.
Just look closely and you will see.

If the big hand is on the *six*
And the little hand passed the *two*
We'll know that it's *two thirty*.
This isn't that hard to do!

The big hand is on the *six*
And the little hand passed the *eight*.
Did you say *eight thirty*?
You are really doing great!

Take a look at this clock
And tell me what time it is.
Did you say *eleven thirty*?
Wow, you really are a math whiz!

Here's a cute ladybug clock.
What time does it tell?
If you said *five thirty*
You are doing really swell!

Thank you, little mouse
For helping us to learn
How to tell time.
But now it is our turn.

We will practice telling time
Each and every day.
When someone asks, "What time is it?"
We'll know just what to say!

Telling Time

Children are so excited when they finally learn how to tell time, especially when they connect it to important times in their lives. *"We eat dinner at six o'clock." "My bedtime is eight thirty."* As in all math skills, there are steps that most children progress through when telling time. First they learn to tell time to the hour and then to the half hour. Sometimes the next step is to learn to tell time in fifteen minute increments, five minute increments, and finally to the minute. Later they will also learn about *elapsed time*, determining what time it will be one hour from now, etc. Eventually they will learn phrases such as *half past, quarter to,* etc. Telling time really involves a lot of skills, including being able to count by fives. Children are exposed to digital clocks everywhere they look but aren't always familiar with analog clocks. Sometimes getting them their very own watch is all the motivation they need to learn how to read an analog clock. I know some people who have covered their digital clocks to provide even more practice telling time. And in the end, that's how children will learn to tell time … practice, practice, practice!

Enrichment Activities

It's About Time

Materials needed:

Two sets of cards (one with clock faces, the other with the matching times) – you can find some great clock clip art online

There are so many things you can do with these telling time cards including …
- ♥ play *Go Fish* with a partner, matching the clock time to the time card
- ♥ play *Memory* individually (place cards in equal rows, face down – turn over two cards at a time and try to match cards)
- ♥ play *Mix-n-Match* (children walk around the room quietly, handing their card to whomever they pass – at a given signal they stop, look at their card, and quietly find the person with their matching card – they stand back-to-back once they have found their match)

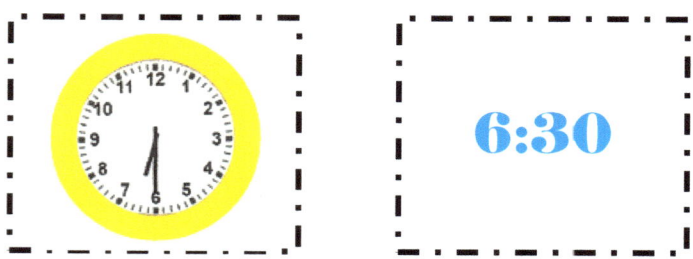

What Time Is It?

Materials needed:

variety of paper clock faces
strips of construction paper (to wrap around children's wrists)
scissors and glue
recording sheet with class names and a spot to write each child's time

How to Play

Children make their own wristwatch, using one clock face of their choice, and place it on their wrist. Using their recording sheet, they walk around the room, find a partner and ask, "Can you please tell me the time?" (or some such phrase). Their partner can only answer by **showing** their watch (no words). The child will write down the time by their partner's name (on their recording sheet) and then find a new partner. Play continues until they have found all the times in the room. You can correct these papers as a whole group or with partners.

Snack Time

Materials needed:

Tick Tock Telling Time (or another "telling time" book)
cookies decorated like clock faces (each with a different time)

This activity is part of my *Story-n-Snack* time, where a parent or other adult from our school or community, reads a story to the children and brings in a snack related to the theme. One mother brought in these wonderful sugar cookies that she and her child had frosted like clock faces! Each cookie had a different time on it. After listening to the story, the children went back to their seats and had to tell us the time on their cookie before they could eat it! Yum!!

You might want to cover the times in the illustrations (with Post It notes) when you first read the book aloud to encourage even more participation and practice!

ABOUT THE AUTHOR

Kathleen Stone is a National Board Certified educator and is currently teaching second grade. *Tick Tock Telling Time* is her fifteenth children's book. Born and raised in Washington State, she and her husband Gary live in the Olympia area. When not teaching, Kathleen can often be found quilting, sitting by the lake reading, and enjoying time with her family (especially her grandchildren)!

Math is all around us
No matter where you turn
Open your mind to the wonders of math
And all that you can learn

www.ingramcontent.com/pod-product-compliance
Lightning Source LLC
Chambersburg PA
CBHW040750200526
45159CB00025B/1827